DNA and the POWER
of the
DOUBLE HELIX

by Mari Biro

For Julian, Michael and Diana

DNA and the POWER
Of the
DOUBLE HELIX

by Mari Biro

In the year 3000, Virus Aegis, an alien of submicroscopic proportions, was traveling in the Universe, following his departure from a distant galaxy. Suddenly, he was sucked into the zygote of the uterus of a female human on Planet Earth.

"GRRRR", he growled, "wrong place!" He could not see a thing, it was so dark inside!"

From there, he slid out through a tiny hole and entered the human's bloodstream. He was rushed at top speed to a gigantic chamber in a thumping heart. Once inside he shot through a maze of blood cells searching for a perfect target.

He stopped when he found what he was looking for. Yep, this will be the first cell that will undergo the Metamorphosis. "Think of it, this female human, made up of trillions of cells!"

Virus Aegis had been chosen by those of his kind to become the sole heir to the throne that was believed to have been stolen by their greatest enemy, the Chromosomes.

Virus Aegis pierced through the wall of the cell membrane. There they were, all forty-six of them. "They call themselves the DNA, the show- offs", he frowned. "Hah, they are nothing but a bunch of ladders with rungs on a double helix collapsed into helpless bundles."

"So, this is what you look like!", cried Virus Aegis. "Your nucleotide chains that form the sides of your ladders will not deliver secret codes to the cells much longer! I'll make sure of that!"

Virus Aegis analyzed the pattern of the secret language of the DNA. It was obvious that the code was determined by the way in which the chemical *bars were combined. He realized that his mission was very simple. All he had to do was change the code! If he succeeded, the Chromosomes would no longer be able to instruct the manufacture of proteins that were made up of amino acids contained in the cell.*

"Hee, hee, hee", Virus Aegis giggled with glee.
"Think of it, instead of brown hair, purple hair for
this human!" "Imagine that! Those loser DNA's

won't be able to produce the proteins they want if I confuse the heck out of them!"

He waited for mitosis to carry out his plan. This would be the perfect chance because that is the time when the Chromosomes duplicate their set of instructions. The ladders widen and the rungs divide in the middle. Each half ladder will stay with its half daughter cell and form a complete ladder and cell again.

Finally, the enemy began the process of mitosis. Virus Aegis struck with his mighty staff and messed up all the codes while he replicated over and over again. The result was total chaos.

Cell after cell became a jumble of confused information. The female human became gravely ill with ovarian cancer.

Months passed. One day, thousands of submicroscopic robots filled the human's bloodstream. They rushed to the rescue of the dying Chromosomes.

With great ease and precision, they rearranged the information on the ladders and restored their order. Through the power of the Double Helix the Chromosomes regained their original codes.

Robocure

Robocure, the leader of the robots grabbed all the Virus Aegis by the scruff of their necks and trapped them into a nerve capsule for Eternity.

The aliens' capture put an end to cancer and AIDS. It also marked the beginning of the era of genetic engineering.

Nine months later, a beautiful baby girl with ...purple hair was born.

The Evil Manipulation of...

Destined

Natural

Acids

Can

Affect

Normal

Cells

Extremely

Rapidly

Also

Intuitively

Destroy

Sapiens

www.ingramcontent.com/pod-product-compliance
Lightning Source LLC
Chambersburg PA
CBHW052046190326
41520CB00002BA/202